Write & Draw Math

Grade 1

by Mary Rosenberg

New York • Toronto • London
Auckland • Sydney • New Delhi • Hong Kong

Editor: Maria L. Chang
Cover design: Tannaz Fassihi
Interior design: Grafica, Inc.
Photos credits © Shutterstock: (FotoDuets), (Hazal Ak).

Scholastic Inc., 557 Broadway, New York, NY 10012
ISBN: 978-1-338-31437-3
Copyright © 2020 by Mary Rosenberg
Published by Scholastic Inc.
All rights reserved. Printed in the U.S.A.
First printing, January 2020.

1 2 3 4 5 6 7 8 9 10   40   25 24 23 22 21 20

Table of Contents

Introduction

Welcome to *Write & Draw Math*! This book offers engaging, ready-to-use activities that help children develop flexible thinking skills about math. Flexible thinking encourages children to "think outside the box," enabling them to apply learning in new ways, brainstorm how to solve word problems, shift to a new strategy when the current one is not working, and understand abstract concepts. They learn to look at the same math problem through different lenses and explore various ways to solve it—employing a concrete method (using manipulatives), a representational method (drawing objects), or an abstract method (writing a math equation). In fact, several activities ask children to tackle a problem using at least two of these methods.

The activities in *Write & Draw Math* are ideal for presenting to the whole class or to a small group. When introducing a new activity, model for children how to do it and then monitor them while they work.

What makes the problems in this book unique is that there is no one correct answer to any of them. This allows children to make and share a variety of math problems and solutions using the same activity page. Since each activity uses unique numbers each time, children can complete the same page multiple times. As they become more adept at manipulating small numbers, challenge children to repeat the same activity using larger numbers. Encourage them to describe their answers using mathematical language—an important part of today's rigorous standards that is often overlooked in math instruction.

Many of the activities require the use of common classroom materials, such as:

- playing cards (Use only the number cards 1 [ace] to 9.)

- number cubes or dice (6-sided)

- paper clips for spinners

- different kinds of counters (teddy-bear counters, linking cubes, pennies, beans)

- pattern blocks

- place-value blocks

Some activities require the use of a spinner, which is provided on the page. To use the spinner, have a child place a paper clip on the spinner and use a sharpened pencil to hold one end of the paper clip in place (see right). The child then flicks a finger to make the paper clip spin.

So, what are you waiting for? Get your first graders started on a fun, learning-filled journey that will build their mathematical reasoning and critical thinking skills!

Math Standards Correlations

The activities in this book meet the following core standards in mathematics.

OPERATIONS & ALGEBRAIC THINKING

A. Represent and solve problems involving addition and subtraction.

OA.A.1 Use addition and subtraction within 20 to solve word problems involving situations of adding to, taking from, putting together, taking apart, and comparing with unknowns in all positions, e.g., by using objects, drawings, and equations with a symbol for the unknown number to represent the problem.

OA.A.2 Solve word problems that call for addition of three whole numbers whose sum is less than or equal to 20, e.g., by using objects, drawings, and equations with a symbol for the unknown number to represent the problem.

B. Understand and apply properties of operations and the relationship between addition and subtraction.

OA.B.3 Apply properties of operations as strategies to add and subtract (for example, commutative and associative properties of addition).

OA.B.4 Understand subtraction as an unknown-addend problem.

C. Add and subtract within 20.

OA.C.5 Relate counting to addition and subtraction (e.g., by counting on 2 to add 2).

OA.C.6 Add and subtract within 20, demonstrating fluency for addition and subtraction within 10. Use strategies such as counting on; making ten; decomposing a number leading to a ten; using the relationship between addition and subtraction; and creating equivalent but easier or known sums.

D. Work with addition and subtraction equations.

OA.D.7 Understand the meaning of the equal sign, and determine if equations involving addition and subtraction are true or false.

OA.D.8 Determine the unknown whole number in an addition or subtraction equation relating three whole numbers.

NUMBER & OPERATIONS IN BASE TEN

A. Extend the counting sequence.

NBT.A.1 Count to 120, starting at any number less than 120. In this range, read and write numerals and represent a number of objects with a written numeral.

B. Understand place value.

NBT.B.2 Understand that the two digits of a two-digit number represent amounts of tens and ones.

NBT.B.3 Compare two two-digit numbers based on meanings of the tens and ones digits, recording the results of comparisons with the symbols >, =, and <.

C. Use place value understanding and properties of operations to add and subtract.

NBT.C.4 Add within 100, including adding a two-digit number and a one-digit number, and adding a two-digit number and a multiple of 10, using concrete models or drawings and strategies based on place value, properties of operations, and/or the relationship between addition and subtraction; relate the strategy to a written method and explain the reasoning used. Understand that in adding two-digit numbers, one adds tens and tens, ones and ones; and sometimes it is necessary to compose a ten.

NBT.C.5 Given a two-digit number, mentally find 10 more or 10 less than the number, without having to count; explain the reasoning used.

NBT.C.6 Subtract multiples of 10 in the range 10–90 from multiples of 10 in the range 10–90 (positive or zero differences), using concrete models or drawings and strategies based on place value, properties of operations, and/or the relationship between addition and subtraction; relate the strategy to a written method and explain the reasoning used.

MEASUREMENT & DATA

A. Measure lengths indirectly and by iterating length units.

MD.A.1 Order three objects by length; compare the lengths of two objects indirectly by using a third object.

MD.A.2 Express the length of an object as a whole number of length units, by laying multiple copies of a short object (the length unit) end-to-end; understand that the length measurement of an object is the number of same-size length units that span it with no gaps or overlaps.

B. Tell and write time.

MD.B.3 Tell and write time in hours and half-hours using analog and digital clocks.

C. Represent and interpret data.

MD.C.4 Organize, represent, and interpret data with up to three categories; ask and answer questions about the total number of data points, how many in each category, and how many more or less are in one category than another.

GEOMETRY

A. Reason with shapes and their attributes.

G.A.1 Distinguish between defining attributes (e.g., triangles are closed and three-sided) versus non-defining attributes (e.g., color, orientation, overall size); build and draw shapes to possess defining attributes.

G.A.2 Compose two-dimensional shapes or three-dimensional shapes to create a composite shape, and compose new shapes from the composite shape.

G.A.3 Partition circles and rectangles into two and four equal shares, describe the shares using the words *halves*, *fourths*, and *quarters*, and use the phrases *half of*, *fourth of*, and *quarter of*. Describe the whole as two of or four of the shares. Understand for these examples that decomposing into more equal shares creates smaller shapes.

Name: _____

Miranda has more marbles than Jack.
Together, they have 8 marbles.
How many marbles could Miranda have?
How many marbles could Jack have?

1. Solve the word problem.

_____ + _____ = 8

Miranda has _____ marbles.

Jack has _____ marbles.

2. Tell how you solved the problem.

_____

_____

_____

#2 Name: _____

Billy had 10 buttons. He gave some to Samir.
How many buttons could Billy have left?
How many buttons could Samir have?

1. Solve the word problem.

$$10 - \underline{\hspace{2cm}} = \underline{\hspace{2cm}}$$

Billy could have _____ buttons left.

Samir could have _____ buttons.

2. Tell how you solved this problem.

_____

_____

_____

Write & Draw Math: Grade 1 © Mary Rosenberg, Scholastic Inc.

#3 Name: _____

Kim went to the store. She bought fewer than 20 pieces
of fruit. She bought 3 more apples than oranges.
How many of each kind of fruit could Kim have bought?
How many pieces of fruit in all?

1. Show how you solved this problem. Draw a picture.

2. Fill in the blanks.

_____ apples + _____ oranges = _____ fruits

Kim bought _____ apples.

She bought _____ oranges.

3. Tell about your solution.

_____

_____

_____

#4 Name: _____

1. Roll a number cube. Write down the number.

2. The mystery number is 5 more. What is the mystery number?

3. What is the sum of your number and the mystery number?

4. What is the difference between your number and the mystery number?

5. Write the math problem to show the sum of both numbers.

_____

6. Write the math problem to show the difference between the numbers.

_____

Write & Draw Math: Grade 1 © Mary Rosenberg, Scholastic Inc.

#5 **Name:** _____

1. Write a target number in the box.
   (The target number should be between 10 and 20.)

2. Add three digits together to make the target number.
   Show four different ways to make the target number.

Target Number

Way #1	Way #2
Way #3	**Way #4**

#6 Name: _____

Tomas counted red, green, and yellow jelly beans.
He had 12 jelly beans in all. How many of each color
of jelly bean could Tomas have?

Tomas's Jelly Beans

Red	
Green	
Yellow	

1. Tell how you solved this problem.

_____

_____

2. Write the math problem.

_____

Asha grows flowers in her garden. How many roses, daisies, and tulips does Asha grow? How many flowers in all?

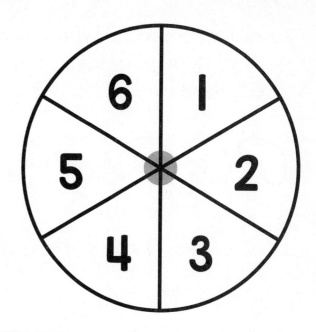

1. Spin the spinner three times. Write a number under each flower.

Roses   **Daisies**   **Tulips**

_____   _____   _____

2. Draw pictures to show how many flowers.

   Roses

   Daisies

   Tulips

3. Write the math problem three different ways.

4. What do you notice about the math problems? Does changing the order of the numbers change the sum?

_____

_____

#8 Name: _____

1. Spin each spinner. Write the numbers below.

_____   _____

2. Use the numbers to write two addition problems. Write two
subtraction problems, too. Draw a picture for each problem.

Addition Problem	**Addition Problem**
Subtraction Problem	**Subtraction Problem**

3. What do the addition and subtraction problems have in common?

_____

_____

Write & Draw Math: Grade 1 © Mary Rosenberg, Scholastic Inc.

#9 **Name:** _____

1. Roll two number cubes.
   Write the numbers in the first two boxes below.

2. Add the two numbers together. Write the sum in the third box.

3. Use the three numbers to write two addition problems.
   Then write two subtraction problems, too.

_____ + _____ = _____

_____ + _____ = _____

_____ – _____ = _____

_____ – _____ = _____

4. What do you notice about the addition and subtraction problems?

_____

_____

_____

#10 Name: _____

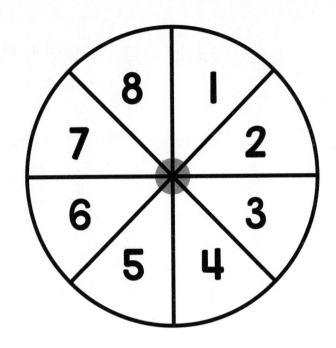

Shane ran 10 laps.
Then he ran some more laps.

1. Spin the spinner. Write how many
laps Shane ran the second time.

Shane's Laps

First time	10 laps
Second time	_____ laps

2. Look at how many laps Shane ran the first time and the second
time. What is the difference between the two numbers?

_____

3. How many laps did Shane run in all? _____

Write & Draw Math: Grade 1 © Mary Rosenberg, Scholastic Inc.

#11 Name: _____

Diego had _____ markers. His friend gave him _____ more markers. Now, Diego has 10 markers.

1. How many markers did Diego have to start with? _____

2. How many markers did his friend give him? _____

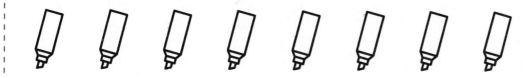

_____ + _____ = 10

Bindi had _____ pennies. She earned _____ more pennies. Now, Bindi has 15 pennies.

3. How many pennies did Bindi have to begin with? _____

4. How many pennies did she earn? _____

_____ + _____ = 15

5. Tell how you solved one of the word problems.

_____

_____

#12 Name: _____

Ben had _____ cupcakes. He sold _____ at the bake sale.
Ben has 8 cupcakes left.

1. How many cupcakes did Ben have? _____

2. How many cupcakes did Ben sell? _____

3. Draw a picture. Write the math problem.

_____ – _____ = 8

Emma has _____ pencils. She sharpened _____ of the
pencils. Now, Emma has 6 pencils left to sharpen.

4. How many pencils does Emma have? _____

5. How many pencils did Emma sharpen? _____

6. Draw a picture and write the math problem.

_____ – _____ = 6

7. Tell how you solved one of the word problems.

_____

_____

Write & Draw Math: Grade 1 © Mary Rosenberg, Scholastic Inc.

#13 **Name:** _____

1. Spin each spinner.

2. Write an addition or subtraction problem.

3. Use the number line to solve each problem.

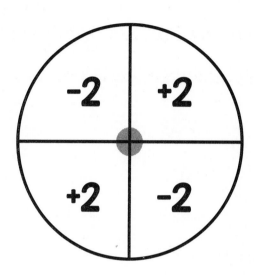

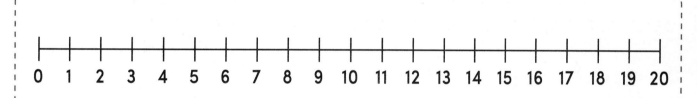

_____ ◯ _____ = _____

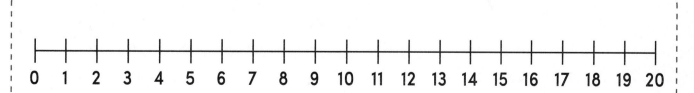

_____ ◯ _____ = _____

#14 Name: _____

1. Roll two number cubes.

2. Decide which strategy to use to add the numbers.

3. Write the addition problem and sum in the correct column.

4. Repeat steps 1 to 3.

Count on 1	Count on 2	Doubles	Doubles + 1	Other
Example: 3 + 1 = 4				

5. Pick one of the addition problems. Tell about the strategy you used to solve it.

_____

_____

Write & Draw Math: Grade 1 © Mary Rosenberg, Scholastic Inc.

#15 **Name:** _____

1. Roll two number cubes.

2. Decide which strategy to use to subtract the numbers.

3. Write the subtraction problem and the difference
    in the correct column.

4. Repeat steps 1 to 3.

Count back 1	Count back 2	Subtract all	Other
	Example: $5 - 2 = 3$		

5. Pick one of the subtraction problems. Tell about the strategy
    you used to solve it.

_____

_____

#16 Name: _____

Ravi thought of a number. Josh thought of a number.
The difference between the two numbers is 7.
What numbers could Ravi and Josh have thought of?

1. Solve the problem.

What is Ravi's number? _____

What is Josh's number? _____

2. Write the subtraction problem.

_____

3. Explain how you solved this problem.

_____

_____

_____

Write & Draw Math: Grade 1 © Mary Rosenberg, Scholastic Inc.

#17 Name: _____

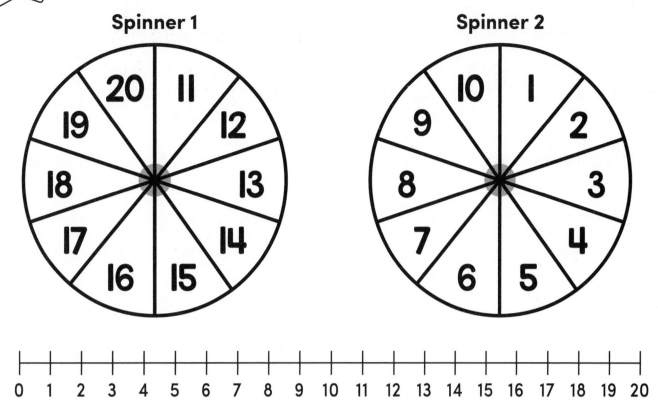

Spinner 1

Spinner 2

1. Spin each spinner. Record the numbers below.

2. Write and solve each subtraction problem.
   Use the number line to help you.

Spinner 1	Spinner 2	Subtraction Problem
		_____ – _____ = _____
		_____ – _____ = _____
		_____ – _____ = _____
		_____ – _____ = _____

#18 Name: _____

1. Roll two number cubes. Write the numbers below.

2. Write an addition and subtraction problem using the two numbers.

3. Cross off the sum or difference on the number line.

4. Repeat steps 1 to 3. Try to use all of the numbers on the number line.

Number Line

Number 1	Number 2	Addition Problem	Subtraction Problem

Write & Draw Math: Grade 1 © Mary Rosenberg, Scholastic Inc.

#19 Name: _____

1. Roll a number cube. Write the number below.

2. Write number sentences with the number as the sum (addition problem) or difference (subtraction problem).

3. Repeat steps 1 and 2. Fill in the chart.

Number Rolled	Equal	Addition Problem	Subtraction Problem
Example: 6	6 = 6	1 + 5 = 6	7 − 1 = 6

#20 Name: _____

1. Solve the given problem.

2. Write an addition or subtraction problem with the same answer.

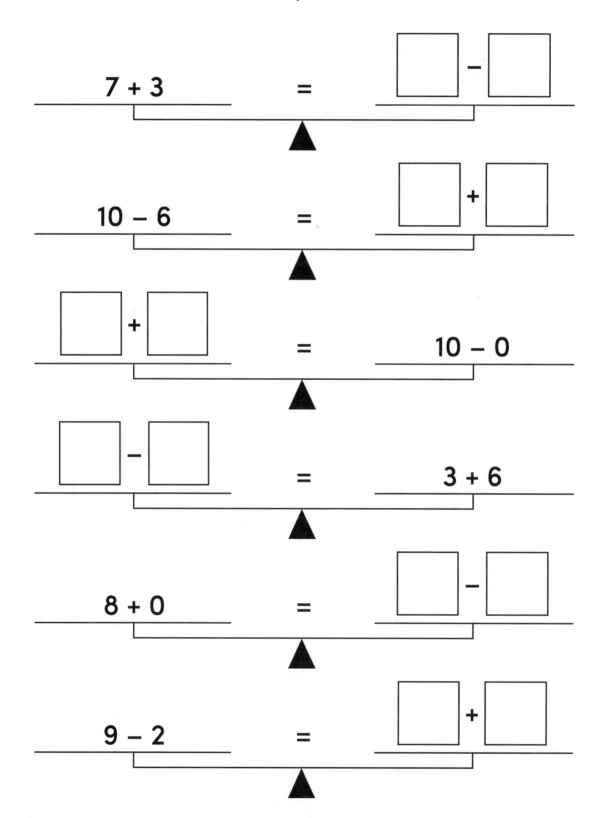

#21 Name: _____

Note to Teacher: Write a target number in the box.

Target Number

1. Write five addition problems.
   The sums should be the same as the target number.

2. Write five subtraction problems.
   The differences should be the same as the target number.

Addition Problems	Subtraction Problems
_____ + _____ = _____	_____ − _____ = _____
_____ = _____ + _____	_____ = _____ − _____
_____ + _____ = _____	_____ − _____ = _____
_____ = _____ + _____	_____ = _____ − _____
_____ + _____ = _____	_____ − _____ = _____

#22 **Name:** _____

1. Roll three number cubes. Write the numbers below.

2. Add the numbers together.

3. Color the sum on the playing board.

4. Repeat steps 1 to 3. Try to color three numbers in a row.

_____ + _____ + _____ = _____

_____ + _____ + _____ = _____

_____ + _____ + _____ = _____

_____ + _____ + _____ = _____

_____ + _____ + _____ = _____

Playing Board

3	4	5	6
7	8	9	10
11	12	13	14
15	16	17	18

Write & Draw Math: Grade 1 © Mary Rosenberg, Scholastic Inc.

#23 Name: _____

1. Start with the first target number.

2. Roll a number cube. Subtract the number.

3. Repeat step 2 two more times. What is the final difference?

**Target
Number**

2	0
−	
−	
−	

1	9
−	
−	
−	

1	8
−	
−	
−	

#24 Name: _____

1. Read each word problem.

2. Spin Spinner 1 to fill in the first number.
Spin Spinner 2 to fill in the second number.

3. Solve the word problem.

Spinner 1

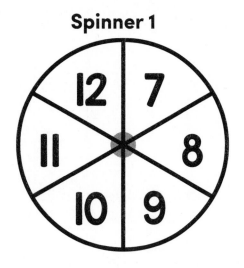

Spinner 2

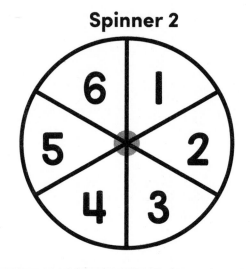

Ray had _____ cards. He gave some to Carla. Ray has _____ cards left. How many cards did Ray give to Carla? Ray gave Carla _____ cards.	Ray had some cards. He gave _____ cards to Carla. Ray has _____ cards left. How many cards did Ray have to begin with? Ray had _____ cards to begin with.
Mavis invited _____ friends to her party. _____ friends came. How many friends did NOT come to the party? _____ friends did not come to the party.	Mavis invited some friends to her party. _____ friends came. _____ friends did NOT come. How many friends did Mavis invite? Mavis invited _____ friends to her party.

#25 Name: _____

1	2	3	4	5	6	7	8	9	10
11	12	13	14	15	16	17	18	19	20
21	22	23	24	25	26	27	28	29	30
31	32	33	34	35	36	37	38	39	40
41	42	43	44	45	46	47	48	49	50
51	52	53	54	55	56	57	58	59	60
61	62	63	64	65	66	67	68	69	70
71	72	73	74	75	76	77	78	79	80
81	82	83	84	85	86	87	88	89	90
91	92	93	94	95	96	97	98	99	100
101	102	103	104	105	106	107	108	109	110
111	112	113	114	115	116	117	118	119	120

What do you notice about the numbers on the 120 chart?

_____

_____

Name: _____

1. Write your favorite digit (0, 1, 2, 3, 4, 5, 6, 7, 8, or 9) in the box.

2. Color all of the numbers with your favorite digit on the 120 chart.

1	2	3	4	5	6	7	8	9	10
11	12	13	14	15	16	17	18	19	20
21	22	23	24	25	26	27	28	29	30
31	32	33	34	35	36	37	38	39	40
41	42	43	44	45	46	47	48	49	50
51	52	53	54	55	56	57	58	59	60
61	62	63	64	65	66	67	68	69	70
71	72	73	74	75	76	77	78	79	80
81	82	83	84	85	86	87	88	89	90
91	92	93	94	95	96	97	98	99	100
101	102	103	104	105	106	107	108	109	110
111	112	113	114	115	116	117	118	119	120

3. What do you notice about the numbers you colored?

_____

_____

#27 **Name:** _____

Note to Teacher: Provide number cubes, a copy of the 120 chart (page 31), and crayons.

1. Roll two number cubes to make a two-digit number. Write the number in the first column.

2. Find the number in the 120 chart. Color it.

3. Write the numbers that are one less and one more. Color them on the 120 chart.

4. Write the numbers that are ten less and ten more. Color them on the 120 chart, too.

5. Repeat steps 1 to 4 two more times.

Two-Digit Numbers	One Less (−1)	One More (+1)	Ten Less (−10)	Ten More (+10)

#28 Name: _____

Note to Teacher: Provide a copy of the 120 chart (page 31).

1. Pick six numbers from the 120 chart.

2. Write the numbers in the middle column.

3. Then write the number that comes before and after each number.

One Less (−1)	Number Picked	One More (+1)

#29 Name: _____

1. Pick a number from Circle A. Then pick a number from Circle B.

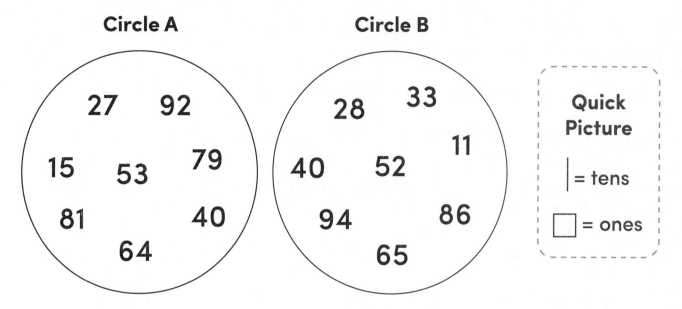

2. Draw a quick picture (using tens and ones) of each number.

Circle A	Circle B

3. Compare the numbers. Use the symbols < (less than),
> (greater than), or = (equal to).

Circle A   Circle B

◯

_____   _____

#30 Name: _____

1. Dip a scoop into a bucket of linking cubes.

2. Snap together each set of ten cubes to make a rod. Leave the "ones" loose.

3. Draw a quick picture of the rods and cubes. Write the number.

4. Repeat steps 1 to 3 three more times.

Quick Picture

| = tens

☐ = ones

There are _____ tens and _____ ones. The number is _____ .	There are _____ tens and _____ ones. The number is _____ .
There are _____ tens and _____ ones. The number is _____ .	There are _____ tens and _____ ones. The number is _____ .

5. Write the numbers in order from least to greatest.

_____ , _____ , _____ , _____

Write & Draw Math: Grade 1 © Mary Rosenberg, Scholastic Inc.

#31 Name: _____

1. Spin the spinner three times.
   Write the digits below.

   _____ , _____ , _____

2. How many two-digit
   numbers can you make
   using the digits above?
   Write the numbers below.
   Draw a quick picture to
   show each number.

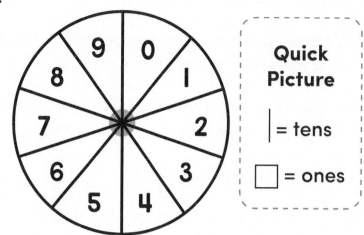

Quick
Picture

| = tens

☐ = ones

Two-Digit Number	Quick Picture

#32 Name: _____

Number of Players: 2

1. Play "Guess the Number." Fill in the number chart below with
   20 consecutive numbers (for example, 1–20, or 31–50).

2. Player 1 writes a mystery number on a piece of paper.
   (The number should come from the number chart.)

3. Player 2 guesses the mystery number. Player 1 answers "higher"
   (if the number is too low) or "lower" (if the number is too high).

4. Based on Player 1's clues, Player 2 crosses off numbers on the number chart.

5. Play continues until Player 2 correctly names the mystery number.

Play this game with a partner. Your teacher will show you how.
Record the numbers below.

Number Chart

Guess	Number	Guess	Number
1		6	
2		7	
3		8	
4		9	
5		10	

The mystery number is _____ .

#33 Name: _____

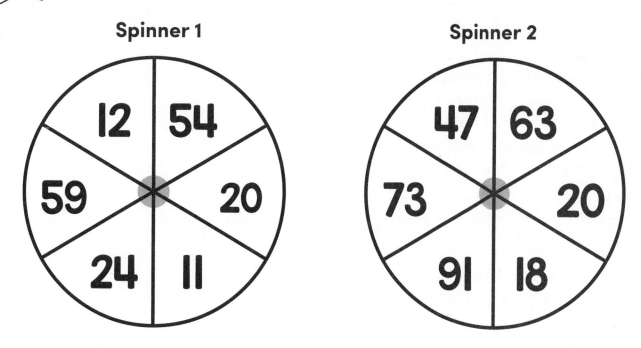

Spinner 1

Spinner 2

1. Spin both spinners. Write the numbers below.

2. Compare the numbers. Use the < (less than), > (greater than), or = (equal) symbols.

3. Repeat steps 1 and 2 a few more times.

Spinner 1	Symbol	Spinner 2

#34 Name: _____

Number of Players: 2 or 3

You'll Need

- different-color crayon for each player
- copy of 120 chart (page 31)

1. Players take turns spinning each spinner.
   Record the symbol and number (for example, > 71).

2. On the 120 chart, color a number that fits the spin (for example, 73).
   Write the number in the chart below.

3. The first player to color three numbers in a row wins the game!

Play this game with classmates. Your teacher will show you how. Record the numbers below.

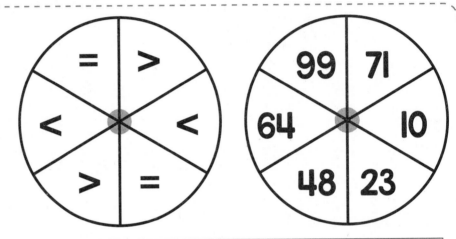

Player 1		Player 2		Player 3	
Spin	Number Selected	Spin	Number Selected	Spin	Number Selected
Example: >71	73	=99	99	<48	42

Write & Draw Math: Grade 1 © Mary Rosenberg, Scholastic Inc.

#35   Name: _____

1. Spin the spinners.

2. Make a number line that begins with the larger number.
Then count on to add the lower number.

_____

3. Write the addition problem.

_____ + _____ = _____

4. Explain how you used a number line to solve the problem.

_____

_____

_____

#**36** Name: _____

Note to Teacher: Provide a copy of the 120 chart (page 31).

1. Spin each spinner.

2. Use the numbers to write an addition or subtraction problem.
   Use a 120 chart to help you solve the problem.

_____ + _____ = _____   _____ − _____ = _____

_____ + _____ = _____   _____ − _____ = _____

_____ + _____ = _____   _____ − _____ = _____

_____ + _____ = _____   _____ − _____ = _____

3. Describe how you used the 120 chart to solve one of the problems.

_____

_____

Write & Draw Math: Grade 1 © Mary Rosenberg, Scholastic Inc.

#37 Name: _____

1. **Read each word problem. Spin the spinner to fill in the missing number.**

2. **Draw a number line (or use a 120 chart) to solve each problem.**

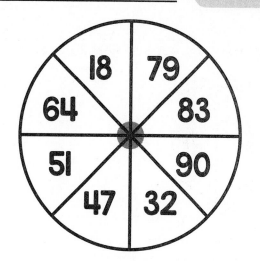

Hanna has _____ cards. She gave 10 to her brother. How many cards does Hanna have now?	Dot has _____ stamps. She added 10 more to her collection. How many stamps does Dot have now?
Hanna has _____ cards.	Dot has _____ stamps.
Kian found _____ pebbles on the beach. He found 10 more pebbles by a log. How many pebbles did Kian find in all?	Matt has _____ toy motorcycles. He gave 10 to his friend. How many toy motorcycles does Matt have left?
Kian found _____ pebbles in all.	Matt has _____ toy motorcycles.

#38 Name: _____

Number of Players: 2

You'll Need
- deck of playing cards
  (Use only the number cards, 1 [ace] to 9.)
- different-color crayon for each player
- copy of 120 chart (page 31)

1. Shuffle the cards. Stack them between players.

2. Player 1 turns over the top two cards to make a two-digit number. She writes the number in the chart below.

3. Next, Player 1 spins the spinner. She adds or takes away 10 from the number and writes the new number in the chart.

4. Player 1 then colors in both numbers on the 120 chart.

   For example: *Say the player turns over 2 and 4 and makes 24. She spins the spinner and lands on "10 more." She colors in 24 and 34 on the 120 chart.*

5. Player 2 takes a turn and repeats steps 2 to 4.

6. Players continue taking turns. The first player to color in four numbers in a row wins.

Play this game with a partner.
Your teacher will show you how.
Record the numbers below.

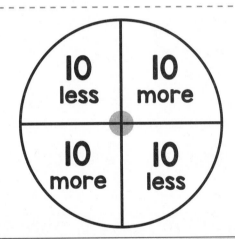

Player 1			Player 2		
Number Made	10 More	10 Less	Number Made	10 More	10 Less

#39 Name: _____

1. Write a starting number in the box. (The number should be a multiple of 10.)

2. Spin the spinner. Use both numbers to make a subtraction problem. Write the subtraction problem below.

3. Use place-value blocks to model the problem. Find the difference.

   For example: *Say the starting number is 90 and you spin 30. Take 9 rods (or tens). Then subtract 3 rods (or tens). Write 90 – 30 = 60.*

4. Repeat steps 2 and 3 a few more times.

Example:

90 – 30 = 60

_____   _____

_____   _____

_____   _____

_____   _____

Write & Draw Math: Grade 1 © Mary Rosenberg, Scholastic Inc.

#40 Name: _____

1. Spin the spinner two times. Use the two numbers to write a subtraction problem.

2. Color the dimes to show the larger number.

3. Cross off the dimes to show the number being subtracted.

4. Write the difference on the line.

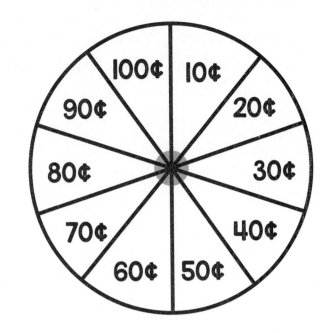

_____ ¢ – _____ ¢ = _____ ¢

_____ ¢ – _____ ¢ = _____ ¢

_____ ¢ – _____ ¢ = _____ ¢

_____ ¢ – _____ ¢ = _____ ¢

Matilda made three trains using cubes.
The trains are different lengths.

1. **Make the trains Matilda could have made.**
   Trace each train.

Train A:
Train B:
Train C:

2. Which train is the longest? _____

3. Which train is the shortest? _____

4. Explain how you figured out which train was the longest.

_____

_____

1. Cut a piece of string or yarn. Tape it to the paper.

2. Measure the length of the string. Use different items.
   The items should touch end-to-end.

3. How many items long is the string?

   _____ paper clips   _____ beans

   _____ counters   _____ pennies

4. Which item did you use the most of? _____

5. Which item did you use the least of? _____

6. Explain.

   _____

   _____

   _____

#43 **Name:** _____

1. Pick two things to measure (for example, pencil and crayon).

2. Trace the length of each thing below.

3. Measure each one. Use two different items (for example, pennies and paper clips).

Thing #1: _____

It is _____ _____ long.
    (number)   (item used to measure)

It is _____ _____ long.
    (number)   (item used to measure)

Thing #2: _____

It is _____ _____ long.
    (number)   (item used to measure)

It is _____ _____ long.
    (number)   (item used to measure)

4. Which item did you use the most of to measure?

_____

Write & Draw Math: Grade 1 © Mary Rosenberg, Scholastic Inc.

#44 Name: _____

1. Pick three items to measure. Draw a picture of each item.

2. For each item, make a guess: How many linking cubes long is it? Record your guess.

3. Then use linking cubes to measure each item. Record how long the item is.

Item #1: _____

I think the item is _____ linking cubes long.

The item is _____ linking cubes long.

Item #2: _____

I think the item is _____ linking cubes long.

The item is _____ linking cubes long.

Item #3: _____

I think the item is _____ linking cubes long.

The item is _____ linking cubes long.

Write & Draw Math: Grade 1 © Mary Rosenberg, Scholastic Inc.

1. Pick three items to measure.

2. Draw a picture of each item.

3. Measure each item. Use paper clips and teddy-bear counters.

Item #1: _____

_____ paper clips long

_____ teddy-bear counters long

Item #2: _____

_____ paper clips long

_____ teddy-bear counters long

Item #3: _____

_____ paper clips long

_____ teddy-bear counters long

#46 Name: _____

1. Find the items below to measure.

2. For each item, make a guess: How many linking cubes long is it?
   Record your guess.

3. Then use linking cubes to measure each item.
   Record how long the item is.

Item to Measure	Guess	Actual
Pencil		
Crayon		
Book		
Eraser		
Scissors		
Shoe		

4. Tell about your most accurate estimate.

_____

_____

_____

#**47** Name: _____

1. Show the time you do each activity. Draw the hands on the clock.

School Starts

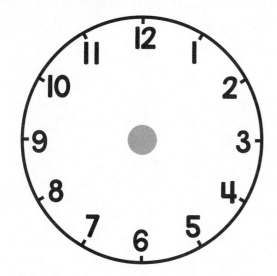

Recess

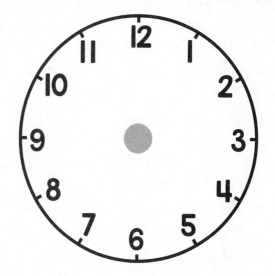

Lunch

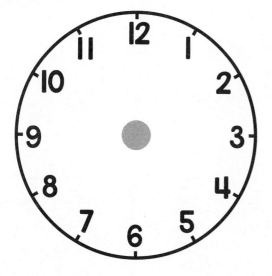

Bedtime

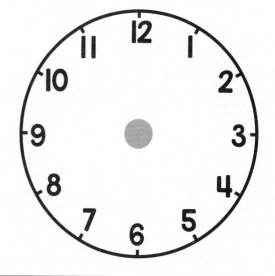

2. Pick one of the activities. Use words and numbers to tell the time two different ways.

_____

_____

#48   Name: _____

Note to Teacher: Set a timer for one minute.

1. How many times can you do each activity in one minute?

Clap your hands: _____ times

Snap your fingers: _____ times

Say the letters of the alphabet: _____ times

Jump up and down: _____ times

2. Make a list of activities you can do in one minute.

3. Make a list of activities you can do in one hour.

4. Compare your list to a classmate's.

Things I Can Do in One Minute	Things I Can Do in One Hour
• _____	• _____
• _____	• _____
• _____	• _____

1. Draw the face of a clock.

2. Count by 5s to add the minutes.

3. Tell what you know about the information shown on a clock.

_____

_____

_____

1. Ask your classmates to name their favorite color. Use tally marks to record their choices.

Red _____

Blue _____

Green _____

2. Make a graph to show the information.

Favorite Colors

10			
9			
8			
7			
6			
5			
4			
3			
2			
1			
	Red	**Blue**	**Green**

Number of Students

Color Choices

3. Write a sentence about the data shown on the graph.

_____

_____

_____

Write & Draw Math: Grade 1 © Mary Rosenberg, Scholastic Inc.

1. Ask your classmates to name their favorite hobby. Use tally marks to record their choices.

Reading _____

Arts and Crafts _____

Sports _____

2. Make a graph to show the information.

Favorite Hobbies

Number of Students	Reading	Arts and Crafts	Sports
10			
9			
8			
7			
6			
5			
4			
3			
2			
1			

Hobbies

3. Write a sentence about the data shown on the graph.

_____

_____

_____

Write & Draw Math: Grade 1 © Mary Rosenberg, Scholastic Inc.

#52 **Name:** _____

1. Select a topic to graph.

Topic: _____

Choice A: _____

Choice B: _____

Choice C: _____

2. Collect your data. Ask classmates to pick one of the choices. Use tally marks to record their choices.

Choice A: _____

Choice B: _____

Choice C: _____

3. Make a graph of the data. Remember to label each category.

Title of Graph: _____

Number of Students			
10			
9			
8			
7			
6			
5			
4			
3			
2			
1			

Choices

Write & Draw Math: Grade 1 © Mary Rosenberg, Scholastic Inc.

#53 Name: _____

1. Draw as many different shapes as possible.

3-Sided Shapes	4-Sided Shapes
5-Sided Shapes	**6-Sided Shapes**

2. Circle your favorite shape. What is the name of the shape?

_____

3. How many sides does it have? _____

4. How many corners does it have? _____

#54 **Name:** _____

1. Make a list or draw pictures of items that fit each category.

Cones	Cylinders	Cubes

2. Circle your favorite solid shape. What is the name of the shape?

_____

3. How many edges does it have? _____

4. How many corners does it have? _____

5. How many faces does it have? _____

Write & Draw Math: Grade 1 © Mary Rosenberg, Scholastic Inc.

#55 Name: _____

1. Pick three different pattern blocks.

2. Combine the pattern blocks to make new and different shapes.

3. Trace the outline of the pattern blocks you used.

Arrangement #1	Arrangement #2
Arrangement #3	Arrangement #4

4. Circle one of the arrangements. Describe it.

_____

_____

5. How many sides and corners does the new shape have? _____

#56 Name: _____

1. Pick three different solid shapes (for example, cube, cone, or sphere).

2. Combine the three shapes to make a new shape. Draw the new shape you made.

3. Describe the new shape you made.

_____

4. What solid shapes did you use to make it?

_____

5. What is the name of your new solid shape?

_____

Write & Draw Math: Grade 1 © Mary Rosenberg, Scholastic Inc.

1. Draw three squares that are the same size.

2. Divide one square into two equal parts. Divide another square into three equal parts, and the last square into four equal parts.

3. Write the fraction for one part of each square.

4. Which fraction is the largest? _____

5. Which fraction is the smallest? _____

6. Order the fractions from smallest to largest.

_____ , _____ , _____

#58 Name: _____

1. Draw a pizza in each box.

2. Share each pizza with a friend or friends.
   Cut each pizza into equal slices to share.

| Share a pizza with one friend. |
| Share a pizza with two friends. |
| Share a pizza with three friends. |

3. **What do you notice about the size of the pizza slices as you share them with more friends?**

_____

_____

Write & Draw Math: Grade 1 © Mary Rosenberg, Scholastic Inc.